CW01429071

Astronom

Beginners

The Introduction Guide To Space,

Cosmos, Galaxies, And Celestial Bodies

Sally R. Ball

Bluesource And Friends

This book is brought to you by Bluesource And Friends, a happy book publishing company.

Our motto is **"Happiness Within Pages"**

We promise to deliver amazing value to readers with our books.

We also appreciate honest book reviews from our readers.

Connect with us on our Facebook page www.facebook.com/bluesourceandfriends and stay tuned to our latest book promotions and free giveaways.

Don't forget to claim your FREE books!

Brain Teasers:

https://tinyurl.com/karenbrainteasers

Harry Potter Trivia:

https://tinyurl.com/wizardworldtrivia

Sherlock Puzzle Book (Volume 2)

https://tinyurl.com/Sherlockpuzzlebook2

Also check out our best seller book
"67 Lateral Thinking Puzzles"

https://tinyurl.com/thinkingandriddles

Table of Contents

Astronomy For Beginners

Astronomy For Beginners

Conclusion

Introduction

Many of you who look up the sky during the day or at night have always wondered about the bodies you see with your naked eyes or by the help of a scientific instrument. Astronomy is the study of heavenly objects. Examples of these objects include galaxies, planets, and comets. Astronomy also involves the study of the universe's unique occurrence that can only be witnessed from the outer part of the Earth's atmosphere. An example of this phenomenon is the occurrence of cosmic radiation.

After reading this book, you will be able to understand the physics, meteorology, evolutionary, chemistry, and movement of celestial bodies together with the development and formation of the Earth's universe.

You will be able to understand the formation of stars that occur in regions of dense gas and tiny solid particles. This is referred to as molecular giants' clouds. It is a very fundamental understanding of the stellar evolution of the stars.

After reading the book, you will understand and learn the discovery of gravity and its relation to the laws of motion in the universe. It is believed that the study of laws of motion and study of gravity was primarily focused on predicting the positioning of the planets, moon, and the sun.

The study carried out on objects in the universe is broad and include, in order of increasing distance, the stars that make up the Milky Way galaxy, the solar system, and more distant galaxies.

Astronomy is inherently more observational rather than an elemental study of science. All measurements are performed at a greater distance from the object of interest, with no control of quantities such as chemical composition, pressure, or temperature. You will also understand the study of the solar system with relation to the gravitational attraction that holds the planets in their elliptical orbits around the sun.

Astronomy For Beginners

An early study of the universe was done through the naked eyes. This method led to the categorization of the celestial bodies and assigned constellations. Constellation has been a very important navigational tool since the beginning of the world. Various disciplines of Astronomy will also be discussed.

Examples of such disciplines include:
- Astrophysics
- Galactic astronomy
- Galaxy Formation
- Cosmology
- Astrometry
- Extragalactic astronomy
- Stellar astronomy
- Planetary sciences
- Astrobiology
- Formation of stars

Chapter 1

Gravitational Force

This is the force of attraction between bodies that have mass. It is proportional to the mass of the object and inversely proportional to the square of the distance between the two objects.

Gravity is a universal force of attraction between all matters in the universe. It determines the motion and structure of every object in the universe. Life without gravity would mean that there would be no planets or stars including the Earth. Planets and stars are formed inside a huge gas cloud and tiny solid particles. The giant cloud of dust is known as a nebula. Planets and stars are formed when the particles in this cloud are pulled towards each other by the gravitational force. So without gravity, there would be no formation of stars and planets, meaning there would be no life.

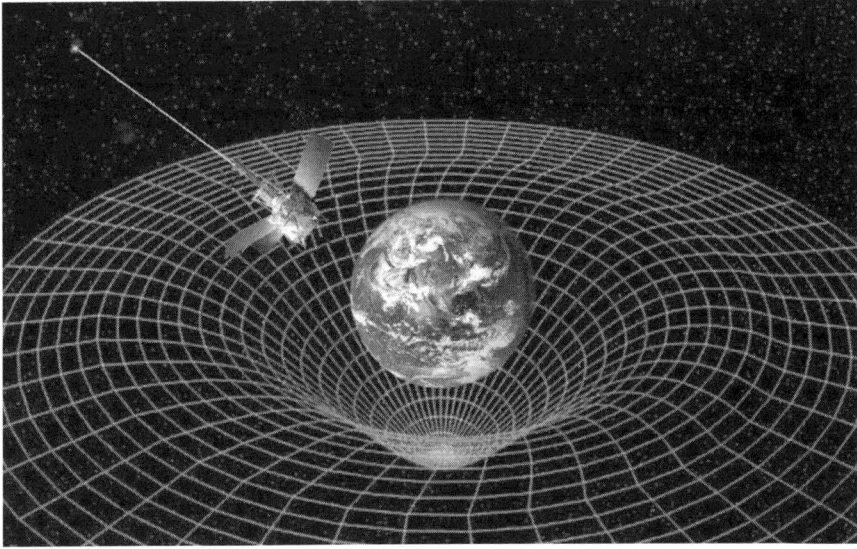

Gravitational force provides the force that maintains the Earth on its orbit around the Sun and the Moon on its orbit around the Earth. Gravity also plays a vital role in holding the atmosphere to the Earth and keeps everything from floating off into space.

Gravity as a Universal Force of Attraction

It is a universal force that tends to pull all matter towards the center of the earth. This force of attraction depends on the mass of the two objects and the distance between the objects. The greater the mass of an object, the stronger the force of gravity; this explains why an object with a greater mass hits the ground quickly and with a big impact. Doubling the mass of one object will lead to a doubling of the force of gravity between the two objects. When you double the mass of the two objects interacting, then the gravitational pull between the two objects will be four times stronger. Therefore, we can conclude that the strength of the force of gravity is proportional to the mass of the first object times the mass of the other object.

Distance and Gravity

When two objects are far apart from one another, the weaker the attraction forces between them. This concept can be defined using the inverse square law. The law states that the gravitational force between two objects decreases with the square of the distance. This distance must be measured from the center and nowhere else. For larger objects such as a planet, the distance considered in the inverse square law must be measured from the Earth's central point. The force of gravity between the Earth and an object always pulls the object toward the center of the Earth. This can be explained by the fact that every piece of the Earth attracts every piece of an object. When all the pieces are brought together, all the forces of attraction add up to result in a larger force pulling them towards the center of the Earth.

Weight in Relation to Gravitational Force

Weight is defined as a measure of the gravity pulling objects to the center of the Earth. The weight of a body will depend on its mass for the same body. Do not assume weight is the same as mass.

The weight of the Earth depends on the mass of the Earth. Mass is a measure of the quantity of matter in an object. In relation to the weight and mass of a person and the gravitational pull of the Earth, when you travel from one place to another, your mass will not change, but your weight would change depending on how far you went from the center of the Earth. The only scientific way for a person to change their weight is by them moving closer or farther from the Earth's center.

In the solar system, there is a tendency curvature nature of the space or the space-time, as the two are inextricably linked. This is due to the Sun, and to a little extent, to Jupiter since it's the second in terms of mass. Since the two bodies themselves moves around but very slightly in the case of the Sun. They result in a dynamic system in which the curvature carries continually as large bodies move around space-time. Mass dictates curvature and space-time dictate the movement of the large bodies.

The Reasons Why the Moon Doesn't Crash into the Earth

Does the Earth pull on the moon due to the force of gravity? Why does the moon not get pulled into the earth and crash?

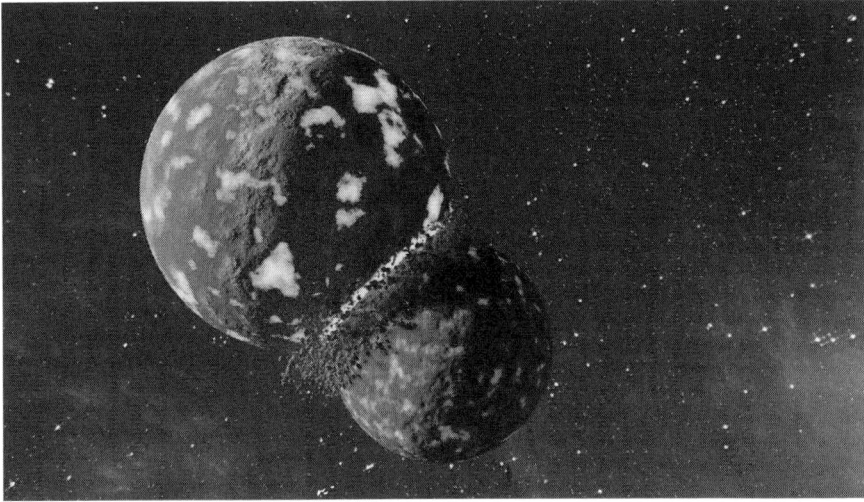

As you had seen in this chapter earlier, gravity is the force of attraction between objects that have mass.

Some of the properties of Gravity include:

- The greater the mass of the objects the greater the gravitational force.

- When you increase the distance between the two objects, the force of gravity between the two objects decreases.

- The force of attraction of one object when interacting with another object is always in the direction of the other object.

- The gravitational force of attraction depends on the mass of both of the objects in the interaction.

Forces

Imagine you have a force on the moon. What do forces do to an object? Yes, it is right to say that forces change the direction of a moving object. Force moves in the same direction as the speed of the moving object.

Supposing an object is moving to the left, then the force pushing the object moves in the same direction. In this case, the force acting on the object will make it move at a higher speed, hence, increasing its velocity.

The force pushing in the opposite direction as the speed of the moving object

In this case, the force acting on the moving object will be acting in the right direction while the object is moving in the left direction. In this case, the force will slow down the moving object decreasing its velocity.

The force pushing the object acts perpendicular to the moving object

This is referred to as a sideways force. In this scenario, the object does not speed up, and it does not slow down. The resulting reaction makes the object turn instead of moving at a constant speed. For the force to cause the sideways movement of the object, the force would have to point in a different direction as the object turns.

If the moon was moving in a perfect circle, the gravitational force would always be acting in the sideways direction and would just make it change its direction. Due to this gravitational force acting sideways, the moon does not speed up and it does not slow down. Hence, the moon will have a constant speed on its orbit, resulting in a constant force of gravity to the center of the Earth. The gravitational force holds up the moon; it's steady position making it not come down crashing into the Earth.

Chapter 2

Understanding The Universe

Universe Evolution

The most commonly accepted theory on the universe origin and universe evolution is the Big Bang model. The model shows how the Earth began with an incredibly dense hot point. Big Bang is explained in detail in the following steps:

1. How it started. It is suggested there was an appearance of space everywhere in the universe from the key experimental conclusions on the cosmic microwave radiation containing the glow of light and radiations left after.

2. First Growth Spurt. There was a burst of expansion that led to the Earth growing exponentially in size at

least ninety times of its original size. During the expansion, the Earth cooled and became less dense. As the fluctuations ended, the process of the universe growing continued, but now, the rate of growth was slower. This led to a formation of matter as the universe cooled.

3. Too Hot to Shine. During the first three minutes of the Universe expansion, chemical light elements were formed. During the process of expansion of the Universe temperatures had cooled down, and the protons bombard with neutrons to form a compound called deuterium which is a hydrogen isotope. The formation of deuterium and its combination in large amounts led to the formation of Helium. High levels of heat from the creation of helium made it intensely too hot for light to shine. The breaking up of atoms into dense opaque protons, electrons, and neutrons of plasma that led to the disintegration of light like a fog was possible by the help of enough force.

4. Let There be Light. There was cooling of the matter to a state well enough for the electrons to interact with the nuclei and combine together to form neutral atoms. As a result of absorption of these electrons, the universe became transparent, allowing light through. Rays of sunlight released at this particular period were able to be detected in the present day in the form of ray radiation from the cosmic microwave radiation.

5. Emergence from Cosmic Dark Ages. Also known as the era of re-ionization. Large particles of gas fell down from the onset of first Galaxies and stars. The light that was emitted from the energetic events, as the Ultraviolet rays cleared out and destroyed most of the surrounding hydrogen gas, was neutral in nature. Re-ionization of the particles plus the clearing out of the foggy hydrogen gas made the universe become a bit more transparent than it was to ultraviolet light initially.

6. Creation of more stars and galaxies.

7. Birth of the Solar System. The Sun is among the more than 100 billion celestial bodies in our Milky Way Galaxy. The sun and other solar system bodies or celestial bodies were created from a huge rotating cloud of gas and tiny solid particles known as a nebula. The effect of gravity on the nebula led to its collapse, hence flattening the nebula into a disk-like object. After the collapse of the huge cloud, most of the materials were pulled toward the center to form the sun.

8. Stuff that Was Invisible in the Universe. At this stage, it was observed that stars at various locations in the galaxies had variation in speed. Stars at the outskirts of the galaxy would orbit more slowly than stars at the center of the galaxy. This movement of the stars in different velocities led to the formation of a mysterious and invisible mass that is called a Dark Matter.

9. Expanding and Accelerating Universe. This was a discovery that was made by the observational use of a Telescope. Astronomer Hubble discovered that the Universe is not stationary, but rather expanding at some rate.

String Theory

This is a unified theory of the Universe that encompasses all the other theories in relation to Evolution of the Universe, including the Big Bang theory. String theory borrows ideas from fundamentals such as forces, particles, manifestations, and interactions as part of one framework.

In-laws of nature, String theory explains remarkably how many similarities there are between unrelated phenomena. In Newton's law, how two massive bodies gravitate, is similar to how electrically-charged particles attract or repel.

At the starting point of this theory is the idea of particles that can be modeled into a one-dimensional object known as a string. It explains how a string propagates through space. In

String theory, a fraction is formed due to the vibrational state of the vibrating strings. A fraction of the mechanical quantum particle brings with it gravitational force, hence, the assumption that the string theory is a branch of the theory on quantum gravity.

String theory attempts to bring together the four forces in the universe namely:

- Electromagnetic force
- Weak nuclear force
- Strong nuclear forces
- Gravity

String theory explains how these forces interacted together in early evolution times. It also explains how particles, either fermions or bosoms, can interact to form a strongly bonded connection.

String theory also brings about another extra dimension in the universe apart from the usual three space dimensions.

The extra dimensions are six. They are curled up to incredibly small sizes that we never perceived to be there.

Chapter 3

Constellation

This is an area in the sky with defined boundaries; all stars and other celestial bodies within that boundary are considered to be part of the constellation. Stars that form patterns in the sky at night are also known as a constellation. In total, we have 88 recognized constellations. Some constellations are visible in the southern sphere while some are only visible in the northern sphere. This phenomenon of stars grouping in the sky at night can be viewed all year round, but some are seasonal and can only be viewed at a certain period of the year.

History of Constellation

For many years ago, mankind has walked the Earth and may have recognized specific patterns formed by celestial bodies such as the stars in the sky. European paintings and carvings that date 10,000 years back bare the marks of star formation and can be familiar to you even today.

Star Charts

Astronomy For Beginners

For easier understanding and to discover various constellations, you will need to first understand the star chart. This gives you a snapshot in your mind of what you will see at night in the sky. The charts may seem rather baffling and confusing at first look, but they are actually easy to understand and simple to use.

The charts are designed in such a way that, sometimes, you may see that the East and West may seem to be on the wrong end, but you need to hold the chart above your head. By doing so, you will realize that the East and West are actually in the right position. The outer edge of the chart shows the horizon; hence, the stars are from the edge the higher they will be formed in the sky.

Holding the chart above your head shows that the center of the chart will show the stars and constellation that will be directly overhead. It will also be helpful for you to find your bearing which will assist you in finding Polaris. Note that the star will always point north.

Astronomy For Beginners

Once a person has found their bearing, they can start searching for the constellation and the objects they contain.

Selected Constellation

1. **Cancer**. This is seen on the southern and northern hemisphere. Cancer constellation is viewed from the late period of autumn to spring in the northern sphere. Cancer will be viewed in the southern hemisphere in the months of summer and autumn. Note that this constellation will appear upside down during this period of the year. Cancer is one of the dimmest constellations. This is due to the fact that it is made up of quite faint stars. The brightest star in this constellation is the Altar star, which is 500 less luminous than the Sun. The main stars in the cancer constellation are as follows: Iota Cancri, Asellus Borealis, Assellus Australis, Acubens, and Altarf.

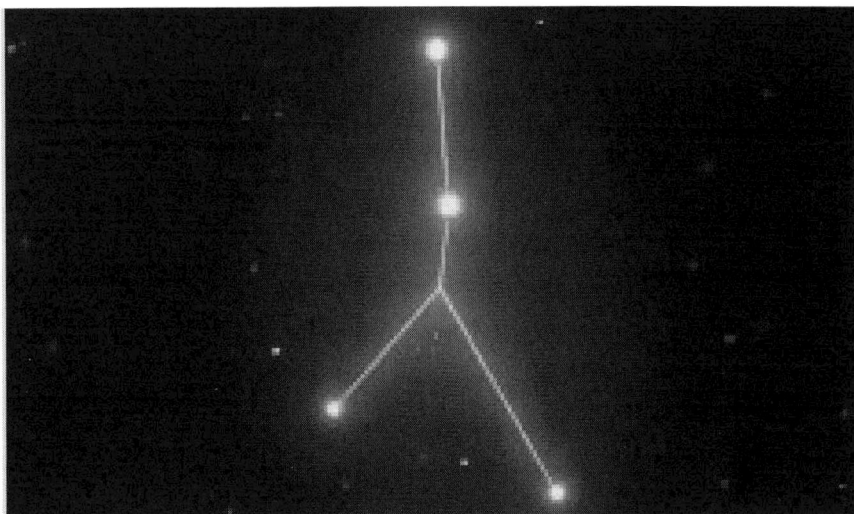

2. **Capricornus Constellation.** Capricornus is observed in the southern and northern spheres. In the northern sphere, it is visible between the month of July and November. In the southern hemisphere, it will be visible from early winter to late spring. Capricornus constellation will be viewed upside down on the southern hemisphere in relation to the northern hemisphere. For you to be able to locate this constellation, you will have to look for a small triangle that forms the head of the constellation. Capricorn is a Latin word for a horned goat. The main stars in the Capricornus constellation include Deneb Algedi,

Nashira, Secunda Giedi, Prima Giedi, and Dabih.

3. **Gemini Constellation**. From Greek mythology, Gemini constellation represents the twin brothers Pollux and Castor. It is visible in the northern hemisphere and most of the southern hemisphere. Gemini can be viewed in the southern hemisphere in the months of summer, while in the northern hemisphere; it is viewed from winter to spring. Gemini appears upside down in the southern hemisphere. The appearance of the Gemini will show two stars at the top of the constellation. This represents the head of the two twins called Pollux and Castor. Pollux and Castor are the brightest stars in the constellation. Main stars in the Gemini constellation include Pollux, Castor, Wasat, Mekbuda, Alzirr, Alhena, Propus, Tejat Posterior, and Mebsuta.

4. **Orion Constellation**. This is one of the most common constellations in the sky at night. It is visible both in the northern and southern Hemisphere. Orion can be viewed in the Northern sphere from the last months of the autumn period to the early months of the spring season. In the Southern sphere, Orion will be seen in the summer months. Orion will be viewed as being upside down in the Southern sphere. The Orion constellation majorly consists of bright supergiant stars. The most observed is the red supergiant Betelgeuse that is found on the shoulder of Orion constellation. Betelgeuse is the biggest star in the constellation with 500 times the diameter of the Sun. The brightest star in the Orion constellation is a star known as Rigel, which is 4,000 times as bright as the Sun. The main stars in the Orion Constellation include the Betelgeuse, Alnitak, Saiph, Alnilam, Rigel, Mintaka, Bellatrix, and Meissa.

5. **Libra Constellation**. It is a dim and a small constellation without the presence of bright stars. It is usually observed in both the southern and northern spheres. Libra is visible in the northern hemisphere between July and April. Libra is visible in the southern hemisphere in the autumn and winter months. It is one of the oldest configurations that

represent a non-living object. The main stars in the Libra constellation include Zuben Elschemali, Zuben Elgenubi, and Brachium.

6. **Virgo Constellation**. This is the second largest constellation in the universe. Virgo is visible in the southern and northern hemispheres. Virgo can be viewed in the northern sphere between July and March. In the southern sphere, Virgo configuration can be seen in the autumn season and winter seasons. Virgo appears upside down in the southern hemisphere. It is densely occupied with several galaxies. The clusters of a galaxy are positioned within the boundaries of the constellation. Virgo consists of the biggest galaxies which are visible through an amateur telescope. Virgo consists of the following main stars Vindemiatrix, Heze, Auva, Spica, Porrima, Zaniah, and Zavijah.

7. **Scorpius Constellation**. Scorpius constellation is a large and bright constellation that is visible in the southern hemisphere. It can be seen in the northern hemisphere in the months of July to August. Scorpius can be seen in the southern hemisphere from the month of March to October. It appears to lie at the center of the Milky Way, making it look like a faint band stretching across the sky. Antares, which is among the stars in the sky that emits a very high intensity of light rays at night, appears at the heart of the Scorpius constellation. Along with many other

bright stars, they belong to a group that consists of bright stars. In this group, the stars were formed around the same period and are from the same region of the Ebola dark cloud. The main stars that form the Scorpius constellation include Shaula, Lesath, Sargas, Antares, Alniyat, Dschubba, and Graffias.

Formation of New Stars

Stars are formed as a result of a dense cloud of gas around the arms of a spiral galaxy. Specific atoms of hydrogen fall with increasing energy and speed toward the nucleus of the cloud under the effect of the star's gravitational force. Due to intense energy, the gas is heated, and continued heating increases the temperature to about 20 million degrees Fahrenheit. A constant series of nuclear reactions begin due to the increased temperature on the hydrogen atoms that causes the star to ignite and burn. The beginning of these reactions marks the formation process of the star.

Collapsing of a Star

Continued burning of hydrogen gas in a star leads to the exhaustion of the gas supply of the star. The first signs of aging stars are reddening and swelling of the outer regions of the stars. A star that is swollen due to aging is referred to as a red giant star. When a star cannot generate enough pressure at its center of gravity, it starts collapsing; this is caused by the exhaustion of fuel. A star collapses due to its own weight. When a star finally collapses, it will generate a violent explosion, blowing up its internal composition into space. The material of the collapsing star will mix together with the primeval hydrogen in the universe. It is out of this mixture that other new stars can be formed.

Chapter 4

Galaxy

This is a system of stellar remains, dust, interstellar gas, stars, and dark material that have been gravitationally bound. Galaxies are composed of stars, gas, and dust that are held together by the gravitational pull. Every galaxy is named according to its shape, making it simple to identify the type of galaxy one is looking for. Galaxies are fascinating because each galaxy has its own unique groupings of black holes, stars, planets, neutron stars, moons, nebulae, dark matter, asteroids, and comets.

The Earth is among the planets in the solar system which is found in the Milky Way Galaxy. The movement of the galaxies suggests that, at one point, they were all together at the same point.

Galaxy Classification

Barred Spiral Galaxy

This type of galaxy has a straight line of bright stars that lie along the nucleus of the bulge and extend into the disk of the galaxy. Inside the bright bulge, there is less activity since it contains old red stars. At the elongated arms of the galaxy is where there is much activity. For example, the formation of stars occurs there.

Elliptical Galaxy

You can recognize this galaxy by its long spherical shape and its lack of a bulge at the center of the galaxy. The galaxy does not have a nucleus at the center. Even with no nucleus at the center, it is still brighter and becomes less bright toward the outer edge of the galaxy. The celestial bodies are evenly spread throughout the galaxy. The galaxy can be long and round forming a cigar shape. In elliptical galaxies, there is less activity, because they contain old stars of low mass due to lack of dust and gas needed for the formation of new stars.

A Galaxy that Is Irregular in Shape

As the name suggests, these are galaxies that have no definite shape, but like other galaxies, they are moving constantly away from the middle of the universe due to gravitational force.

Spiral Galaxy

Astronomy For Beginners

A common feature that helps you to identify a spiral galaxy in a simple way is the disk and a bulging feature. This galaxy contains a nucleus at the center. The nucleus has a spherically-shaped bulge that holds old stars that do not have any gas and dust.

In a spiral galaxy, the disk forms a circular shape. The arm is where the galaxy originates from the disk, and it is where most of the activities of the galaxy take place.

Milky Way

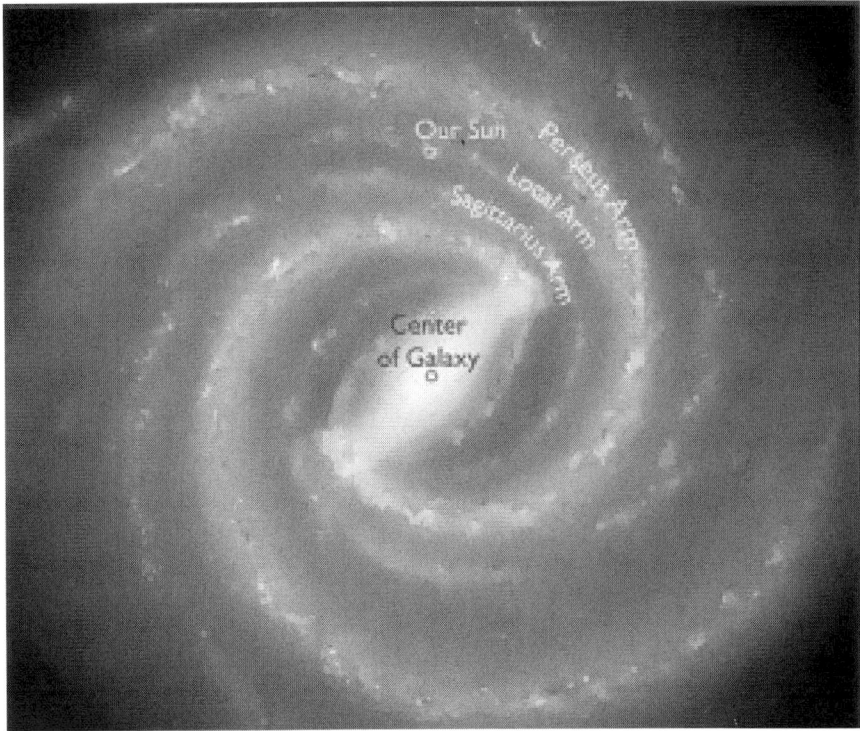

This is considered as one of the spiral-barred galaxies. After the occurrence of the Big Bang, the Universe cooled for a while, and all the gases were uniformly and evenly spread. Due to some minimal irregularities, the gas condensed into larger particles enough for heating up and eventually triggered the nuclear fusion that is responsible for powering the stars. The stars began attracting each other, forming bigger clusters of stars gravitationally. An example of this group is known as the globular cluster. This type of cluster is found at the Milky Way galaxy, which is dated to the very early days of the universe evolution.

One interesting phenomenon about the Milky Way is that it generates at least seven stars or more every year. The Milky Way galaxy is referred to as a cannibal galaxy because it swallows up a smaller galaxy during its formation. At this very moment, it's is believed that the Milky Way galaxy is gobbling up another galaxy of the stellar cluster known as the Canis Major Dwarf galaxy.

Old stars are found distributed in the Galactic halo of the Milky Way galaxy, meaning it is likely that the galaxy had a spherical shape to start out its formation. The disk of the Milky Way contains new bright stars that have been formed recently.

Structure of the Milky Way galaxy

The structure of the Milky Way is fairly the same as a large spiral system. This formation consists of the following six separate parts.

1. Center Bulge. The bulge surrounds the nucleus. The bulge is a formation of stars that are nearly spherical in shape; that is made up of population 2 stars. A globular cluster of stars is found in this region in the galaxy.

2. The Nucleus. This is the center of the celestial body; this is where remarkable objects are found in the galaxy. The nucleus is a black hole that is huge in size, and that has been surrounded by an accretion disk of high temperature.

3. The Disk. This is a part of the galaxy that cannot be seen easily with naked eyes. The disk extends from the nucleus to nearly 75,000 light years. It is thought that the disk is the underlying body of stars upon which the arms of the galaxy are formed. It has a thickness that is roughly one-fifth of its diameter. Besides, it is

made up of different compounds that have different thicknesses.

4. Spiral Arms. Due to the blocking of interstellar tiny solid particles and the interior positioning of the solar system, the spiral structure is very difficult to observe optically with naked eyes. In the Milky Way galaxy, there are three spiral arms.

5. Massive Halo. There is a massive halo at the exterior, to the whole part that is visible. This is caused by the effect of the outside rotation curvature of the galaxy. The halo extends considerably 100,000 light-years from the middle, and its quantity of matter is as many times larger than the quantity of matter of the galaxy summed together.

6. Spherical Property. It is a large space that is below and above the disk of the Milky Way that contains an extension of the central bulge. A sphere-shaped component matches the elliptical galaxy and follows

the law of how the density of an object varies with distance from the center of the object.

Solar System

This is the gravitationally-bound planetary system of the Sun and the celestial bodies that orbit the sun either directly or indirectly.

The Composition of the Solar System

At the heart of the solar system is the Sun that influences the motion of all the other bodies through its gravitational force. The Sun is the most common, largest, and most massive star of the solar system. It makes up to 99.8% of the solar system. It is also called the engine of the system. Even though the planet's gravitational pull may move the center of the solar system away from the Sun's center, it stays deeply encroached within the core of the Sun.

Planets are known to be the second biggest celestial bodies present in the solar system. They are eight in number, and they can be divided into two types: the Terrestrial planets that include Mars, Venus, Earth, and Mercury, and the gas giant planets that include Neptune, Jupiter, Uranus, and the Darrin. All these eight planets revolve around the sun on their orbits in an elliptical, roughly circular orbit in almost the exact plane. In the system, there are no planet orbits that are exactly circular in shape or in the same plane as that of the rotation of the Sun.

Since Jupiter is closest to the Sun, its orbit is nearest to the plane and circular, while that of Pluto moves away from both the plane and from the circularity. Asteroids and comets are also members of the solar system. Asteroids are composed of rocks while comets are majorly made up of volatile compounds and ice compounds. Minor objects like the comets and the asteroids may be found anywhere in the Solar system, with orbits ranging from elliptical to circular.

In the Solar system, there are belts that may surround one or more planets. First, we have the asteroid belt that is found between Jupiter and Mars, and it is entirely formed by asteroids only. Secondly, we have the Kuiper belt, which is found outside the orbit of Neptune, and it encompasses a large area for the Sun. Kuiper belts mainly consist of the Comets. Thirdly, the Oort cloud is found in the solar system. It is assumed to extend out to roughly 365 light-days away from the Sun.

The Organization of the Solar System

It can be categorized into three main regions:

1. Inner systems. It is made up of terrestrial planets, their moons, comets, and close-orbiting asteroids.

They are objects that are exclusively composed of rocks with no atmosphere. The boundary of the inner system has the main asteroid belt that separates it from the outer system.

2. Near Outer System. This is composed of gas giant planets and their moons, comets, and asteroids that orbit between the Kuiper belt and the main asteroid belt. Objects in this system may contain rocks, liquid gas, and ice as major components. The near outer system is defined by the boundary at the orbit of Neptune.

3. Far Outer System. It consists of the Kuiper belt, the Oort cloud, ice Pluto planet, and comets that orbit between the belt and the clouds. Compounds may be made up of rocks. However, in most cases, they are mainly composed of ice.

Solar System Boundary

The boundary is clearly defined in two ways. First, the gravitational boundary can be described as the point at which the celestial bodies no longer orbit the Sun. It includes the Oort cloud. Secondly, the boundary declared by the heliopause of the solar system.

The Planets

These are astronomical bodies orbiting a star that are large enough to be surrounded by their own gravitational pull that should not be enough to cause thermonuclear fusion.

Here are some interesting facts you can learn about the planets.

- There are no moons on the planet called Venus.
- Planet Mars had a thicker atmosphere once in the past.

- Did you know that Jupiter is a great comet catcher?
- Little-to-no discoveries have been made about the age of Saturn's rings.
- Uranus is stormier than people think.
- Neptune has supersonic winds.

- The Earth's magnetic field can be seen at work during light showers.

- Mercury. This is the smallest planet. It is also the fastest planet, orbiting around the Sun every 88 days. The surface of Mercury during the day is heated at high intensity by the sun's rays, reaching a stumbling 450 degrees Celsius. At night, the temperatures of the surface of this planet drop to hundreds of degrees below freezing points. Mercury lacks an atmosphere to absorb the impact of meteors; hence, it has pockmarked, crater-like features on its surface similar to those of the Moon.

- Venus. The thickness of the atmosphere of Venus is very big, making it trap heat in the runaway greenhouse effect. Therefore, it is the hottest planet in the solar system. Venus spins slowly and in the opposite direction as the other planets.

- Planet Earth. The Earth is the only planet known that is inhabited by living organisms and can support life. It is the third largest planet in the solar system and the only planet with water on its surface. A large portion of the planet is covered with oceans. The atmosphere of the Earth is full of nitrogen and oxygen gases that are responsible for the sustainment of human life and other living organisms.

- Mars. This planet is dusty and cold. It is a desert with a very thin layer of the atmosphere.

- Jupiter. This is the largest planet. It is more than twice as massive as other planets in the solar system.

- Saturn. This is a planet that is made up of rings, which are a result of a combination of ice and rock particles.

- Uranus. This is a giant planet with the equator nearly at right angles to its orbit. Uranus is approximately the same size as Neptune. It contains methane in its atmosphere. Methane gives the planet the blue-green tint.

- Neptune. Neptune is known for its very strong winds. The winds are faster than the velocity of sound. Furthermore, it has a rocky core.

- Pluto. This is known as a dwarf planet due to its small natural size. It is said that Pluto is smaller in size than the Earth's moon. A Pluto orbit is inside the orbit of Neptune, and thus it is carried way out beyond that orbit. Pluto is very cold and rocky with a very ephemeral atmosphere.

Dust and Gas in Space

Most people wonder where stars come from. Stars must die because they eventually exhaust their nuclear fuel. New stars are formed to replace the ones that die. For new stars to come into existence, we need raw materials to make the new stars. During the existence of stars, they eject mass throughout their life span; this mass of material goes somewhere when it is blown away from the surface of the stars.

In recent discoveries, it was concluded that vast quantities of raw materials (atoms or molecule) of tiny solid dust and gas are found between the stars. These tiny solid dust particles and gas are referred to as the interstellar matter. Most of the interstellar matter is concentrated into giant clouds; each of these clouds is known as a **nebula**.

These clouds do not last for a long time. Instead, they are like clouds on Earth that are in constant motion, shifting and merging with one another. They either grow or disperse.

Some of these clouds become dense and massive enough to collapse due to their own force of gravity. By doing so, the clouds form new stars. When stars die and collapse, they release some matter into the interstellar space. The matter released is used in the formation of new clouds. The cycle is repeated over and over again.

99% of the material between the stars is in the form of gases; it consists of individual molecules or atoms. The most common element of the gas in the interstellar space is helium and hydrogen. One percent of the interstellar space consists of solid frozen particles made up of many atoms and molecules that are known as interstellar dust.

Interstellar dusts - These are tiny solid particles in the interstellar space consisting of a core of rock-like materials.

Interstellar medium - It consists of the dust and gas found between the stars in the galaxy.

Nebula - This is a cloud of interstellar gas or dust forming a cloud that is often seen glowing with visible light or infrared light.

Interstellar Gas

Hydrogen is the main component of the interstellar gas compound. A region is characterized according to whether the hydrogen component is ionized or neutral. Interstellar gases also contain other components besides hydrogen. Most of these elements are ionized in the vicinity of hot stars, which then capture electrons and emit light.

Interstellar Dusts

They can be detected when:

- They block light from stars behind the dust particles.
- They make distant stars look fainter and redder.
- They scatter the light from nearby stars.

Interstellar dusts are found throughout the plane of the Milky Way galaxy. They are of the same size and have the same wavelength of light and consist of rocky substances that are either sand-like or soot-like.

Chapter 5

Comets and Meteors

Meteors

This is a visible path that a meteoroid passes through the Earth's atmosphere. It is commonly known as a shooting star. After the meteoroid enters the Earth's atmosphere, it becomes bright and visible because of the heat emitted by the friction between the atmosphere and the meteoroid.

When a meteoroid survives a transit process through the Earth's atmosphere, it rests on the Earth's surface resulting in the formation of an object called a meteorite. In the beginning, during its entry into the upper part of the atmosphere, a trail of ionized ions is formed. This is where the molecules in the upper part of the atmosphere are ionized by the incoming meteor. Ionization process takes up to 45 minutes. Ionization trail may be more or less continuously found in the upper atmosphere, since there is a constant entrance of small, sand-sized meteoroids every few seconds.

Meteor Showers

Meteor showers vaporize almost all their material in the Earth's atmosphere leaving behind a bright trail commonly known as a shooting star. Sometimes, there is the occurrence of many meteors at night in the sky. If the number of meteors increases dramatically, the event is known as a meteor shower.

Comets

These are icy celestial bodies that emit dust and gas. Comets contain ice, dust, ammonia, methane, carbon dioxide, and many more elements. From past studies, it is believed that comets might have brought some of the organic molecules and water on Earth. The molecules and water are essential for life on Earth.

Comets are believed to revolve around the Sun and perceived to be present in the Oort cloud. Oort cloud is a specific space found beyond the orbit of Pluto. It is rare for comets to streak through the inner side of the solar system. Some comets are seen once every few centuries. Most people die without seeing a comet, but for those who are lucky to see it, they will not want to forget.

Physical Properties of a Comet

The inner part of a comet consists of tiny solid particles and ice that is coated with dark organic material. Ice is frozen water. A comet can also contain other components such as ammonia, methane, carbon monoxide, and carbon dioxide. The Nucleus of the core of the comet may have a small rocky part. When comets begin to move towards the sun, the frozen water on the surface of the nucleus starts to evaporate into gases due to the heat produced from the sun's rays. After the evaporation, a cloud called a coma is formed. Radiations from the Sun's rays cause the tiny, solid particles to move away from the cloud, creating a string like a tail of dust. Charged particles from the Sun convert gases into ions, forming a tail of charged ions. The tail of comets is shaped by sunlight and solar wind, making the tails to point away from the Sun's direction.

Asteroids and comets seem to be similar in appearance. However, asteroids do not have tails, unlike comets, which have tails.

Orbital Properties of the Comets

They are classified according to the period they take to orbit around the Sun.

Short-period comets need roughly less than 200 years to orbit around the Sun in one complete revolution. Long-period comets take more than 200 years for them to complete one orbital revolution around the Sun. Singleapparition comets are bound not to go around the sun; their orbits move away from the solar system.

Short-period comets are formed from a disk band of icy objects known as the Belt of Kuiper that is found beyond Neptune's orbit. Due to gravitational pull with objects in the outer side of planets, comets are attracted inward, making them very active. Long-period comets come from the nearly spherical Oort cloud. These comets slung inward due to the gravitational attraction of the passing stars.

Artificial Satellites

It is an object that mankind launches in the Earth's orbit with the purpose of carrying out a scientific study or research. Currently, we have thousands of artificial satellites that orbit the Earth. They have been designed for different altitudes, and they are in different sizes depending on the purpose.

Altitude and Sizes of the Satellites

A communication satellite is about seven meters long, consisting of solar panels that extend another fifty meters. One of the largest artificial satellites is the International space station. It consists of a five-bedroom structure that houses the station personnel.

Variations of Altitudes above the Earth's surface

- Low Earth Orbit. It is positioned at about 200-2,000 kilometers. For example, the International space station orbits at 400 kilometers with a speed of 28,000 kilometers per hour. It takes ninety minutes to complete one orbit.

- Geostationary Orbit. It is designed to be 36,000 kilometers above the Earth's surface. It takes one day to complete one revolution that is equivalent to 24 hours. It is mainly used for weather and communication purposes.

- Medium Earth Orbit. It is a type of satellite that is

positioned at about 20,000 kilometers above the Earth's surface. It takes twelve hours to complete one orbital revolution.

Types of Satellites

- Communication Satellite. It is mainly used for internet transmission, and television transmission, too. An example is a D1 satellite, which is in geostationary orbit above the Earth's Equator.

- Weather Satellite. It is mainly used to provide images of cloud patterns, and measure rainfall and temperature for weather prediction patterns. Low Earth Orbit and the geostationary satellites are used depending on the purpose of the satellite.

- Astronomical Satellite. It is mainly used for monitoring and taking images of the space.

- Earth Observation Satellite. It is mainly used for taking images of the Earth. Low Earth Orbit is used since it produces a more detailed image.

- Navigation Satellite. It consists of the Global Positioning System that has 24 satellites positioned at an altitude of 20,000 kilometers above the Earth's surface.

Satellite Design

Basic Parts of a Satellite

1. A powerhouse. Most of the satellites have solar panels that are used to generate electricity for the powering of the station. Some contain batteries to store energy for the satellite in times that the satellite is in the path of the Earth's shadow.

2. Bus. It consists of the total framework of the satellite where all the parts of the satellite are attached.

3. Heat Control System. This satellite is exposed to the high temperatures of the sun. Therefore, there is a need for the satellite to reflect and radiate heat.

4. Communication System. It is a very integral part of the satellite to send signals and receive data to ground stations. That's why many satellites have curved dishes to act as antennae.

5. Altitude Control System. Have you ever asked yourself why and how the satellite dishes point in a certain direction? This is the system that is necessary to keep

the satellite pointing in a particular direction. Gyroscopes are used to change the orientation of the satellite while light sensors are used to determine the direction that the satellite will face.

6. Propulsion System. This system is used to place the satellite in the correct position on the orbit. Once the satellite is on the orbit, it does not need rockets anymore. It requires small rockets known as thrusters that help in small changing of positions on the orbit.

Space Trash

Space trash is the mass from defunct artificially created objects found in space, especially in the Earth's orbit. The trash is made up of old satellites, spent rockets, and fragments from their disintegration and collision.

The size of the space trash ranges from microscopic to obsolete space crafts and rockets that stand many kilometers high, with no functionality in the Earth's orbit.

Risks of Space Trash

Orbital trash travels very fast compared to other orbital bodies in space. In the Low Earth Orbit, the space trash has an average velocity of 21,600 meters per hour. Just try to imagine how a tiny particle could cause havoc moving at that speed.

Causes of Space Trash

The main contributor to space trash is the object break up in the Earth's orbit. These break-ups can be caused by collisions or explosions. An explosion could result from overheated batteries, residual propellants, collisions, and deliberate destruction of the satellite.

Chapter 6

The Eclipses

An eclipse is a partial or complete obscuring of a celestial body by another celestial body. This occurs when the celestial bodies are positioned in one line.

Lunar Eclipse

This phenomenon takes place when the Moon passes behind the Earth directly into the shadow of the Earth. It occurs when the Earth is between the moon and the sun.

There are three categories of the lunar eclipse: penumbra, partial, and total lunar eclipse. Of these, the most spectacular phenomenon is the total lunar eclipse where the shadow of the Earth completely covers the moon. An eclipse does not occur at every full moon.

There are two types of shades that the Earth casts upon the Moon during the lunar eclipse: the Umbra and the penumbra. Penumbra is an outer partial shadow, while the Umbra is a full, dark shadow.

The Moon undergoes all of these stages of the lunar eclipse. This is the initial and final stage of the eclipse where the moon is not noticeable at the penumbra stage. The most spectacular phenomena of a lunar eclipse are the occurrence of the eclipse at the middle of the event. This is when the moon is in the umbral shadow.

1. Umbra stage. Umbra is also known as the total lunar eclipse. This event occurs when the Earth's full shape

is cast on the shadow of the moon. This does not mean that the moon will disappear completely. The moon is cast in the darkness and that makes it possible to miss seeing it if you were not staring at the lunar eclipse as it occurs. The moon is given a dim glow; this is possible because some light rays going through the Earth's atmosphere are evenly scattered and the light is refracted and refocused on the moon. This is the light that reflects on the moon during a total lunar eclipse.

2. Partial lunar eclipse. Any eclipse can be a partial eclipse. Total lunar eclipse experiences the partial bit of an eclipse process in its totality stage. During the occurrence of the partial eclipse, the Sun, the moon, and the Earth are not aligned in a straight line.

3. Penumbra Lunar Eclipse. This phenomenon occurs if the moon is in the Earth's giant outer part of the shadow. It is the least interesting type of eclipse. This type of lunar eclipse is difficult to notice because the outer side of the Earth's penumbra will be pale.

Blood-Red Moon

The moon may appear to turn red or copper in color. This occurs when some of the sunlight rays pass through the Earth's atmosphere; they are bent and directed toward the Moon during the Umbra stage of the eclipse. The red color of the spectrum tends to penetrate through the Earth's atmosphere, compared to the other colors of the spectrum that are blocked and evenly scattered by the Earth's atmosphere. The red color effect casts itself on all the planets sunrises and sunsets.

Solar Eclipse

This phenomenon occurs when the Moon is at the center of the Sun and the Earth. The moon casts its shadow on the Earth's surface. By doing so, the moon obscures the Earth's view of the Sun. This event only occurs when there is a new moon; at this point, the

Sun and Moon are in conjunction as seen from the Earth's position. The total solar eclipse is a very rare occurrence due to its totality, and it can only be viewed where the Moon's umbra shadow touches the Earth's surface.

Types of Solar Eclipses

• Annular Solar Eclipse. It takes place when the disk of the moon is so small to cover the Sun's disk fully. This leaves the Sun's outer edge visible thus forming a red hot ring in the sky. This occurrence takes place if the Moon is near the apogee, while the Moon's umbra touches the surface of the Earth.

• Partial Solar eclipse. This event occurs when the Moon obscures the Sun's disk partially, making the moon to cast a penumbra shadow on the surface of the Earth.

• Total Solar Eclipse. As the name suggests, it takes

place when the Moon's spherical disk completely covers the Sun's disk. It only occurs when the Moon is close to its perigee. A perigee is a point on the Moon's orbit that is closest to the Earth.

• Hybrid eclipse. It is a very rare type of eclipse that occurs if an eclipse is the same and it converts from an annular eclipse to a total Solar eclipse and also from a total Solar eclipse to an annular eclipse. This has to occur along the eclipse's path.

Reasons why there is no Solar Eclipse at every New Moon

- The new moon is close to the lunar nodes. Nodes are specific points in which the plane of the Moon's orbital path is around the Earth and meets with the Earth's orbital plane around the Sun.
- It is also compulsory for the Sun to be near to the two lunar nodes. This will lead to the formation of near perfect or perfect alignment with the Earth and the

Moon. The alignment takes place every six months, and it lasts for an average duration of around 34.5 days.

Wormholes in Space

It is a way of traveling through space-time that connects two distant points in the space of the universe. Examples of wormholes are in movies such as *Interstellar*, where characters are seen using wormholes as portals to distant parts of the galaxy. One of the demerits of the wormhole is that there is no observational evidence that they exist and there is no empirical proof that they do not exist.

For wormholes to be stable, they must be assisted by some kind of exotic matter. An exotic matter is a matter that has a negative mass rather than a regular mass that has a positive value.

Black Holes in Relation to Wormholes

Is there another formula to create wormholes? Theoretically, it would be possible to create wormholes by the assistance of Black holes. The process is known as the Einstein-Rosen Bridge. This is the formation of wormholes by the immense warping of Space-time by the effects of a black hole. Specifically, it involves the use of a black hole called the Schwarzschild black hole. One of the properties of this black hole is that it has a static amount of mass, does not rotate, and has no electrical charge.

As light falls into the black hole, light passes through a wormhole and it is released out through the other side, through a material known as the white hole. The white hole is similar to the black hole; the difference is that the white hole would suck in the material while the black hole will repel the material.

Is there Any Possibility in the Future for Man to Use Wormhole?

Sidelining the technical aspects of wormholes mechanics,

some other challenges arise in terms of the hard physical truths about these objects. Even if we prove that they exist, it will be challenging for humanity to control or manipulate them. There are also questions about the safety of wormholes. Who knows what to expect inside the wormholes?

The Possibility of the Existence of Aliens

Have you ever asked yourself of possible life in another parallel universe? You should do that. If research of possible aliens living in the universe comes out with nothing substantial about the existence of aliens, there might be a worthy course in checking or researching on a neighboring Universe. In relation to the universe, the theory of the idea of our universe is just one of the many theories about universes.

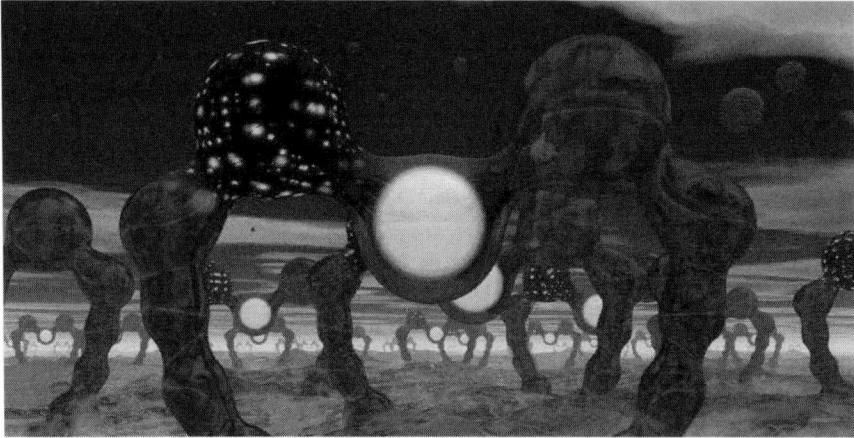

In a new computer simulation that was done to build a new universe under specific conditions, it was discovered that the existence of life might be broader than earlier perceived. The mysterious force of Dark Energy confirmed that there is a high likelihood that there exist aliens in another universe.

Dark Energy

It is an invisible force that is mysteriously thought to exist in the spaces of the universe. You might have thought of gravity pulling things close together; the dark energy flings matter apart. The dark energy is responsible for the expansion of the universe due to its constant and invisible push. The speed at which the universe is expanding gets faster and faster, day by day.

Science suggests that as space is created in the universe, the more the dark energy appears to fill the empty spaces created. According to recent scientific research, it has been discovered that nearly seventy percent of the mass-energy of the universe may be composed of dark energy.

Expanding Universe

It was believed that the gravitational force of attraction slows down the extension of the universe over a specified period. However, when the rate of expansion was measured and calculated it was found that it was accelerating.

The accelerated expansion rate of the universe is driven by a particular repulsive force that is created by quantum inflations in the space in the universe. Studies have shown that dark energy that occupies the spaces in the universe is a type of fundamental force called quintessence. Dark energy fills the spaces in the universe just like a fluid.

Astronomy For Beginners

The Existence of Dark Matter

Dark matter is a matter that does not release light or energy. The Universe contains much matter than what can be seen by our naked eyes. It contains materials called the baryonic that consists of neutrons, electrons, and protons. Dark material may be made of non-baryonic or baryonic matter.

Dark matter is a missing matter that cannot be detected easily and it is made up of baryonic matter. Super-massive black holes could also be part of the dark matter in the spaces of the universe.

Dark matter also consists of neutrino stars, dim brown dwarfs, and white dwarfs. It could have played a dominant role in the missing mass during scientific studies. Dark matter consists of ten-to-a-hundred times the mass of a proton. However, as a result of the weak interaction with regular matter, the dark matter was difficult to detect. It is different from other regular matter. The following experiment can be used to detect unusual material:

- The Alpha Magnetic Spectrometer. This is a very sensitive detector of particles that was installed on the international space station, which has been in operation since 2011. The detector has been able to detect more than a hundred billion cosmic rays.

Chapter 7

Telescoping

Evolution of Telescoping

Even in the early days before the invention of telescopes, people did make an observation of the stars and other celestial bodies. In ancient Egypt, they used a tool called the merkhet. They made an observation on certain stars that passed at different times to measure the time it took to move about in the sky. A similar concept was used by the Greeks. They built instruments that measured the angle of stars with respect to the horizon. Having a track of the angle of the star over time, they were able to calculate a rough path of the star through the sky.

Modern Day Telescope

A telescope is an instrument used by astronomers that makes distant objects appear nearer and larger by gathering and concentrating radiations.

Principals of telescoping

1. Reflecting telescope. It consists of a concave mirror that is used for gathering light rays from the viewed object and further refocusing it onto the eyepiece that is adjustable or a series of lenses. The eyepiece is responsible for the reflection and enlargement of the objects being viewed.

2. Refracting telescope. It is made up of an objective lens. The lenses are set at one end of the tube together with an eyepiece that is adjustable with a series of lenses. The lenses are placed on the other end of the tube to allow it to slide into the first tube. This is the point where the object is enlarged and viewed directly.

The Advantages of Using a Telescope

- You do not need to look at the Earth's atmosphere directly with your eyes. If you need detailed information, you may need a telescope since the Earth's Atmosphere is pretty murky and horrible.

The Disadvantages of Using a Telescope

- One of the main demerits of using a telescope is the hassle of how to operate it for maximum use.
- Most of the modern telescopes are very expensive to purchase.

Conclusion

This book contains up-to-date and detailed information about celestial bodies, space, and galaxies among other topics associated with Astronomy. You will also gain in-depth knowledge about various chemical processes that occur in our universe without our knowledge here on Earth. This book gives interesting information about the different star constellations that we have in the sky. A star chart can be used to view various constellations in the sky at night. Celestial bodies such as comets are rare to witness because they appear once in a few centuries. There are two major types of eclipses, namely Solar and Lunar Eclipses. This book gives a clear distinction between the two types of eclipses. Besides, the book explains clearly how solar and lunar eclipses are formed. An explanation about shooting stars, meteors, and meteoroids is also provided in detail to the reader. The author helps you understand how stars are formed in the universe. An overview of the evolution of telescoping has also been explained in detail.

Connect with us on our Facebook page www.facebook.com/bluesourceandfriends and stay tuned to our latest book promotions and free giveaways.

Printed in Dunstable, United Kingdom